Cover illustration by Jacob Voigt
jacobvoigt.com

ISBN: 978-0-9885821-5-6

WAYPOINT

MAPPING THE FUTURE OF TALENT IN THE AGE OF AUTOMATION

JARED NICHOLS

WITH FOREWORD BY DON W. HUBBLE:
FORMER CHAIRMAN, CEO AND PRESIDENT
OF ANGELICA CORPORATION

THE
JARED NICHOLS GROUP
empowering organizations to shape the future

way·point

/ˈwā,point/

noun

A set of coordinates which define a point in space.

- Waypoints are useful in designing various autonomous missions for quadcopters. Mapping out would be impossible without a possibility to define these physical locations.

CONTENTS

FOREWORD

There can be no doubt that there is a talent and skills shortage in most organizations today, from entry level management to the C-Suite. There are myriad reasons for these shortages. In this book the shortages are identified but, more importantly, actions to diminish (elimination is a most challenging task) the shortages are addressed. This includes recommendations on how to focus upon elements which will effectively improve performance, Executive Foresight™ being one such tool.

The Millennial demographic is receiving what may be termed "excessive attention" in the popular press. However, what is not excessive is the fact that the uniqueness of the psychographic composition of millennials cannot be discounted. Certainly workers today demand a different work-life balance than did preceding generations. The meaning of work, how work is structured, and how recognition is given are significantly different with those workers aged 30 and younger from those who are older than 45. For the group aged between 30 and 45, they tend to behave and react similar to the millennials at times, and at other times more like there older counterparts.

The collective differences between the workforce of today and of a decade or two ago are significant. The desire for more frequent feedback is one such difference; even if the expectations are for mainly positive feedback. The desire to do primarily "meaningful work" is another difference; a job isn't simply a paycheck. The absence of loyalty to an organization is much more prevalent today; the causative factors created both by the worker and the organization. Expectations of having more responsibility at earlier stages in one's career have been heightened with higher earning power as a consequence. The absolute need to continue to learn and to have intellectual curiosity is essential for success in the future. The ability to use the power of Big Data will be a major differentiator between effective and ineffective leaders.

There is no absence of data on virtually any subject today and the ability to access both specific and random databases easily provides a cornucopia of information upon which to make intelligent decisions. This is both good news and bad news. A leader's ability to select disparate pieces of information from many different sources, coalesce these to solve perplexing organizational problems, and create road maps for future strategies and actions plans separates the winners from the losers. (This is no different today than it was in the past with the notable exception that there is infinitely more data available.)

Big Data is truly Big Data. The Nobel Prize-winning economist Herbert Simon warned: "Information consumes the attention of its recipients. Hence a wealth of information creates a poverty of attention." The cliche and practice of "paralysis of analysis" needs to be avoided. Special attention by the reader should be given to the S.T.E.E.P. categories presented in the book, the two most important ones being Social and Economic. The human element can never be ignored nor can the economic impacts of plans and actions. This includes the economics of the organization and the economics of individual workers and groups of workers, the impact of which is not a zero-sum endeavor.

No matter how effective any organization is in utilizing the power of Big Data, people still matter. As stated in the book: "critical thinking, creativity and new ideas are essential for consistency of success." The three P's (Purpose, Passion, and Perseverance) are immensely important. While facts and logic really matter, they often become a bit fuzzy when opinion and emotions come into play. Every leader should attempt to allow some opinions and some emotions to be considered in any decision-making process, the organization is best served when facts and logic prevail.

The reader should pay particular attention to Chapters 5 and 6 to understand the power of "Zeroing In" and the critical importance of "Scenario Building." (Scenario Building should never be a casual exercise!) Younger workers of today are very comfortable with most tools of technology. The more committed workers are also aware of the need to continue to add new tools to their respective tool-kits. Coupling these characteristics with intellectual curiosity predicts individual success. Organizations are attempting to add workers who are "emotionally intelligent, highly organized, excellent communicators, and creative problem solvers." Unfortunately these workers are in short supply and too few organizations have programs in place to add these desirable characteristics to their existing work force. Being aware of this shortfall should be the catalyst to attract and retain those workers committed to intellectual curiosity, continued learning, and self-improvement.

Hiring decisions - often using the power of Big Data - can be most helpful in this regard. Corporate culture can be reinforced and employee retention and productivity can be improved. A word of caution: Diversity is still desirable and a strict reliance on Big Data when attracting and retaining may be counter-productive as a consequence; never forget about the human factor. "Sameness does not always predict the best ultimate results."

Chapter 8 should be of particular interest to those that want to be "better tomorrow than they are today." The pivotal question: "How do we leverage what we now know about the future?" is an excellent beginning. Virtually all of the questions in Chapter 8 need comprehensive attention and reflective thought by the critical reader.

Finally, the recognition that the future changes every day and that the need for personal improvement is not only necessary but also beneficial to the individual and the organization. Those workers that continue to be successful enjoy the process of improvement and revel in the joys of intellectual curiosity.

Don W. Hubble
Former Chairman, CEO and President – Angelica Corporation

ABOUT JARED NICHOLS

Jared Nichols is a deep futures strategist, executive advisor, speaker, and coach. He provides the tools to help leaders and organizations gain competitive advantage, seize new market opportunities, drive in new revenue, and increase profits. As one of the few people in the world to hold a Master's Degree in Strategic Foresight, Jared is sought out by leaders, organizations, and entrepreneurs to help them identify and create their long-term successful future.

Jared's insight and expertise is utilized across a wide variety of sectors and industries from Fortune 500 companies to government municipalities, entrepreneurial start-ups, as well as some of his most recent work in Hollywood with accomplished actors, writers, and producers, helping them reinvent themselves and discover new areas for growth both inside and outside the bounds of their industry. Jared is also the creator of the Executive Foresight™ Program, a unique program specifically designed to teach leaders how to anticipate change, navigate uncertainty, and identify new opportunities for growth in several possible future outcomes.

Jared sits on the Board of Directors for the National Small Business Association with a focus on equipping SME's with the ability to anticipate change and influence public policy in favor of small business growth.

As a highly prolific author, Jared has published over 50 articles and is the author of Rethinking Your Next Quarter (Century): How to Create Continuous Growth and Ensure Future Relevance, Rethinking Reinvention, Leading the 21st Century: The CEO's Guide to Thriving in a Volatile and Uncertain Future, and Four Futures for the 21st Century Non-Profit. Jared is also a musician, composer, competitive cyclist, and trail runner living in Charlotte, North Carolina with his wife and their two sons.

A NOTE FROM THE AUTHOR

The content of this book was conceived and written based on my work with Rod Nease and Andy Hewitt of Sunbelt Rentals, Inc. When we first began working together through the Executive Foresight™ Program, our goal was to gain greater insight into the possible future impacts and opportunities of artificial intelligence, telematics, robotics, and Big Data.

As one of the largest equipment rental companies in North America, the impact that these trends and issues could have on Sunbelt Rentals' ability to grow, attract top talent, and continue to provide superior service to its customers was immense. As we continued to work together, it became clear that our findings had wide sweeping implications that were applicable across several industries and organizations. As such, we decided to take our work together and compile it into the book you are reading now.

Although our work started out with a specific focus on the equipment rental industry and the target customer groups of Sunbelt Rentals, what you will find in the following pages is a roadmap that you can apply to your own organizational strategy for long-term growth.

Jared Nichols
Principal – The Jared Nichols Group

INTRODUCTION

Today's leaders and organizations are faced with unforeseen challenges on a daily basis, struggling with the question of future relevance, growth, and profitability. Unfortunately, the majority of those challenges have been met with antiquated rhetoric, methodology, and systems that were created for a stable and predictable market environment. The fact is we no longer live in a stable and predictable market environment where fluctuations are manageable and disruption is anticipated long before it arrives. The symptoms of instability and uncertainty are manifesting in every major institution, as evidenced in the public's attitude toward these institutions. In the U.S. a 2015 Gallup poll showed that with the exception of the military, the public's confidence in government institutions is staggeringly low. In addition to government, public confidence in financial institutions is at a near record low, coming in at 28%. Faith in our education system at every level remains low amidst poor competency scores compared to the rest of the world., and confidence in organized religion hit an all time low coming in at 42% in 2015.[1] Amidst an atmosphere of instability and uncertainty, it has become increasingly difficult for leaders and organizations to know where to invest their time, energy, and capital.

This uncertainty does not plague the leadership level alone. The question of human relevance to the workforce has become increasingly ambiguous as more companies, governments, and institutions look to automation and robotics to increase productivity, reduce costs, and increase profits. As this trend continues to grow, the implications this will have on the economy, organizations, and communities will be significant to say the least. The good news is that there is still time to take the necessary actions to position individuals, organizations, communities, and entire industries and sectors to thrive. This book is not a theoretical treatise on why we must plan for the long-term future. This need has already been well-established. What we wanted to accomplish here was a way forward, through a "how-to" process that can be applied across sectors. By illustrating the method we used to evaluate our talent and workforce development strategies going forward, we have also created a tool that any organization can use to test their own strategies. We hope that by outlining this process, we can empower other organizations to create their long-term successful futures.

[1]**Gallup, *Confidence in Institutions***

CHAPTER ONE

THE CURRENT TRENDS IN TALENT & WORKFORCE DEVELOPMENT

The nature of work, employment, training, and talent development are areas that are all rapidly evolving. We see the signs of this in four major issues:

1. The talent and skills gap
2. The current leadership vacuum
3. The growing "on-demand" talent market
4. The growing importance of work/life balance

These four trends and issues are currently having a major impact on the way we work and the degree of success we can hope to achieve. To further understand these impacts, we must look more closely at each issue.

THE TALENT AND SKILLS GAP

The education sector is often one of the slowest to adapt to changes in the market and the overall business landscape. This impediment has become glaringly obvious in the growing talent and skills gap, in which the current talent pool does not possess the skills and expertise needed for the jobs available. Public confidence in the value of institutions of higher learning is greatly waning. Only 13% of Americans strongly agree that college graduates are prepared for the workforce.[2]

Not only are educational institutions failing to adequately train, but once in the workforce, on-the-job training is greatly lagging in its ability to keep pace with the rapidly changing environment. A recent survey finds that less than half of employees feel their companies provide adequate technological training. In addition, only 23% of executives surveyed offered training and development as an employee benefit.[3] Company loyalty is less of a given in today's workplace, and in response employers hesitate to invest in employees the training and development so greatly needed.

Additionally, the methods of training and development are often outdated. The old models rely too heavily on increasing competencies, which, while important, do not address the overall shifts occurring in the global economy. What we know is important, but an equal investment must be made in the way we think. To remain relevant in today's workforce, employees need the critical thinking skills to make sense of their changing environments and the ability and autonomy to make important decisions in realtime. Often employees are needed to not just be skilled in multiple disciplines (i.e. science, business, and technology), but also understand how these disciplines intersect and impact each other.[4]

Perhaps the greatest untapped resource in the struggle to close the widening skills gap is our unprecedented access to information and ideas. In fact, in a recent survey, Millennials stated that they much preferred online learning (68%), while only 4% preferred university classes.[5] The shift away from the hierarchical dissemination of information through formal institutions to a network-based collaborative sharing of skills and knowledge provides potentially unlimited opportunities for growth and development if properly utilized. This brings us back to teaching our workforce how to make sense of the rapidly changing world around them. Understanding how to interpret the impact and importance of the information available to them may be an employee's greatest asset in the 21st century.

THE LEADERSHIP VACUUM

Employers are not cultivating leaders within their organizations anywhere near the rate needed. A recent survey cited lack of leadership as executives' second greatest obstacle to meeting their workforce objectives.[6] It is predicted that succession planning will be a top priority for employers this year as a large portion of Baby Boomers retire.[7] The recent recession has taken resources away from leadership development while making employers more hesitant to hand the reins over to younger, less experienced employees. This problem has only been exacerbated by the growing healthcare expense burden on employers.

Despite the mischaracterization that Millennials and younger employees do not desire leadership positions, 91% of young people surveyed stated they aspire to a position of leadership; however their concept of leadership is that of a transformational leader, rather than the autocratic leadership style of the Boomer generation.[8] They also prefer less hierarchy within an organization and greater fluidity. The nature of leadership is changing from an emphasis on a single person or role to an emphasis on a network of collective leadership and collaboration.[9] Such trends beg the question if some leadership positions may not be filled until organizational structures are better adapted to their changing environment. The current leadership models might lack the capacity to develop the type of leadership needed to navigate an increasingly complex and volatile landscape.

Employers are trying multiple strategies to fill the leadership vacuum: asking senior executives to delay retirement, hiring retiring leaders back for contract work, and implementing mentorships to foster leadership among younger employees. Unfortunately, these strategies will only go so far if organizations do not rethink their current leadership models within the broader context of their changing environment. The formal leadership role and title is becoming less important than creating opportunities for leadership to emerge through employee actions.

THE GROWING "ON-DEMAND" MARKET

Since the recession, companies have continued to streamline operations and shed non-essential personnel. At the same time, there has been a growing need to remain agile enough to quickly meet rapidly changing needs. Increasingly, companies are meeting this need with contract or freelance workers. Companies are understandably hesitant to create permanent positions in such a rapidly changing environment, especially when it can be more cost-effective to tap the on-demand talent pool. New research has found that contract or temporary employment jobs has grown much faster over the last decade than conventional jobs.[10]

Even furthering this trend are workers' shifting attitudes towards freelance or contract work. Many are seeing it as an opportunity to maintain greater control over their careers with greater flexibility for work/life balance. Workers now place greater value on discretionary time, autonomy, and mobility. The erosion of trust between companies and employees in terms of longterm stability and loyalty has also had more Americans entering the gig economy. It is projected that by 2020, 40% of the American workforce will be freelance.[11] Whatever the cause, it is clear that contingent workers have become increasingly vital to many organizations' success.

THE NEW WORK/LIFE BALANCE

The tools and devices we use to work and live have become smarter, so too are the ways in which we balance these two areas.The jobs that once required clocking in to perform repetitive tasks, such as assembly line work, or tedious computing, such as generating routine reports, are now being relegated to computers and robotic technologies. The jobs that are left require complex, creative, and adaptive thinking. Smart workplaces are discovering that these attributes are fostered in an environment that promotes work/life balance. Technology has enabled us to very effectively work remotely. Workforce mobile platforms are continuing to grow and increase in functionality. We now realize that productivity isn't necessarily tied to time spent at the office. Increased flexibility has allowed us to work when and where we are most effective, while technology has boosted our ability to collaborate and our access to needed information.

Technology may have provided the avenue for increased work/life balance, but the economy has created the drive. In the fallout from the recession, as so many lost their jobs and investments, many saw it as an opportunity to reprioritize. Many young people, who watched parents lose everything after devoting their lives to a job or a company, decided to invest in their personal lives rather than in a specific job or company. Many now seek jobs that fit their preferred lifestyles rather than adapting their lifestyles to a specific job. However, the desire for greater fulfillment extends from workers' personal lives to their careers. Workers are now more focused on their careers and what they can achieve and contribute to society rather than just keeping a specific job. For instance, employees fear not having the latest and most relevant skills and technological expertise twice as much as they fear being laid off.[12] Workers are seeking organizations that provide opportunities for growth and development that will help them stay relevant and lead to greater fulfillment in both their professional and personal lives.

[2]**Busteed, *America's "No Confidence" Vote on College Grad's Work Readiness.***
[3]**SAP and Oxford Economics, *Workforce 2020.***
[4]**CSBI, *Life Sciences Workforce Trends Report 2014, 4.***
[5]**WorkplaceTrends, *The Millennial Leadership Survey.***
[6]**SAP and Oxford Economics, *Workforce 2020.***
[7]**Schawbel, *10 Workforce Trends for 2016.***
[8]**WorkplaceTrends, *The Millennial Leadership Survey***
[9]**Petrie, *Four Trends for the Future of Leadership Development, 6.***
[10]**Irwin, *With 'Gigs' Instead of Jobs, Workers Bear New Burdens.***
[11]**Schawbel, *10 Workforce Trends for 2016.***
[12]**SAP and Oxford Economics, *Workforce 2020.***

CHAPTER TWO

APPLYING THE FUTURE TO THE PRESENT DAY

While there's no way to know for sure what the future will bring, the major emerging trends of today can provide some pretty good insights to where we're headed. Among the trends that are likely to have the greatest impact on our future are:

1. Our increasing reliance on technology
2. The utilization of Big Data
3. The rise of intelligent machines

We can already see the impact of each of these in our lives today, but that impact is likely only to grow as we move into the future.

INCREASED RELIANCE ON TECHNOLOGY

Today, we rely on technology to perform all sorts of tasks that we performed ourselves just 10 to 20 years ago. Information that we gathered from printed materials or television, such as maps, news, and weather reports, are a mere click away. We hold meetings virtually and can in many cases work from wherever we please. We shop and order gifts or food in a matter of moments and know that the items will reach their destination without us ever having to leave our chairs. We have so much power at our fingertips, we simply just have to ask our devices for whatever we need at that moment.

However, as machines and computers become increasingly intelligent, there will come a time when we will not even have to ask. Our devices will be able to anticipate our needs and desires before they even arise. Consider Amazon's move to purchase a patent on anticipatory shipping. It might be hard to accept that a machine will know what you want before you do, but the more we use our devices, the more familiar they become with our patterns of behavior, our interests, with whom we network, how we communicate, and even how we think.

The scope of smart technology will expand into new areas of our lives. We see this already in our homes with products that allow you to monitor and control systems in your home remotely. Our machinery, such as automobiles, computers, medical equipment, and heavy machinery, will continue to become more intuitive and easier to operate, requiring less skill and thought on our part. As technology continues to do things for us that we once did for ourselves, our reliance as we move into the future is certain only to increase.

THE DATA-DRIVEN FUTURE

Although we have yet to truly experience the power of Big Data, its promise looms large. As Big Data continues to amass information in the Digital Age, its proponents believe it holds the answers to many of our biggest problems. If we could only learn how to better utilize the unprecedented amount of information out there, we could work smarter, faster, and find better solutions.

Many industries have already begun to do this. Engineering, design, and urban planning firms are already capturing and using data to plan smarter buildings, campuses, and communities. The data combined with smarter computer-driven processes allows computer applications to test thousands of design plans in a matter of seconds and perform the majority of a design project's decision-making process.[13] On a more controversial note, many police departments already use crime prediction applications driven by data about past crimes to determine where to patrol during their shifts.[14] As facial recognition software becomes more widely used and links up to our online data, concerns about online surveillance, privacy, profiling, and the use of data are sure to grow.

While the construction industry is notoriously slow to adapt to emerging trends, it potentially has much to gain from the use of Big Data. Reducing inefficiencies, cutting costs, and shortening projects could all be achieved through capturing better data and learning how to use it. New technologies, such as telematics, have attempted to pull the industry out of the digital Stone Age, but its current capabilities only go so far and knowing what to do with the information once it is acquired is another problem. Big Data by itself does not provide the insights needed to streamline the construction process from start to finish. Knowing what data to collect and how to use it is a dilemma that all industries trying to harness the power of Big Data currently face. This dilemma is further complicated by an inability to share data and communicate across multiple systems.

RISE OF THE INTELLIGENT MACHINES

Perhaps the answer to the Big Data dilemma lies in the creation of smarter machines, machines that are capable of analyzing data in a way that will eliminate the need for human decision-making. Experts have speculated about the potential of artificial intelligence for decades, while research and development continue to bring us closer to its reality. We can all remember when IBM's Watson defeated Jeopardy champions in 2011. Watson's current endeavors include data analytics, to help average businesses make sense of Big Data. Many other large corporations are getting into the AI game, like Google's work to use machine learning and AI to better understand human queries when processing searches; and a separate application designed to understand and reply to emails on your behalf.[15] We have all heard of Google's autonomous vehicle development; however, many major car manufacturers (Toyota, Honda, Mercedes-Benz, BMW, and Audi to name a few) are also investing billions in the race to make driverless cars a reality.[16]

We already see the impact of automation in the heavy machinery industry. Semi-autonomous and fully autonomous mining trucks have been in use for a few years now. Caterpillars' semi-autonomous dozers can be controlled remotely, requiring a fraction of the manpower and making operations safer for humans. Caterpillar's and Komatsu's fully autonomous mining haulage systems merely require monitoring, which can be done from virtually anywhere. Looking into the future, A.I. is certain to further disrupt the construction industry, which has suffered declining rates of productivity in recent decades.[17] More highly specialized equipment fills a need for skilled labor, while potentially reducing human error and inefficiencies. From Termes (Harvard's self-organizing robots modeled after termite colonies), to brick-laying robots, to the construction site data-gathering capabilities of unmanned aerial vehicles, further construction site automation is on the horizon.[18, 19]

Of course many other industries will feel the impact of automation as computers become more highly skilled at doing our jobs. A much referenced study out of Oxford has predicted that 47 percent of current U.S. jobs will be automated over the next decade or two, from insurance underwriters to tax preparers.[20] The jobs that were once considered safe from automation due to a certain level of skill are now at risk due to advances in technology. The study also finds that for the time being, the tasks least susceptible to computerization require creative and social intelligence. Job loss is not the only potential downside of A.I. Many leading scientists in the field are concerned that further development be done responsibly, ensuring A.I. remains safe and beneficial for humans.

WHERE DO WE GO FROM HERE?

Thinking about what these emerging trends mean for us and our future can be very overwhelming. What challenges and opportunities might these trends present? How will our lives and livelihoods be affected? Strategic planning often goes by the wayside when individuals and organizations don't have the right toolbox for sifting through the data, asking the right questions, imagining alternative futures, and creating the best possible solutions. The following section gives an overview of the method we used to tackle these issues.

[13]Barista, *The Big Data Revolution: How Data Driven Design is Transforming Product Planning.*
[14]Huet, *Server and Protect: Predictive Policing Firm Predpol Promises to Map Crime Before It Happens.*
[15]Emspak, *Google's New AI System Could Be 'Machine Learning' Breakthrough.*
[16]CB Insights, *30 Corporations Working on Autonomous Vehicles.*
[17]Skibniewski and Golparvar-Fard, *Toward A Science of Autonomy for Physical Systems: Construction.*
[18]Hegeman, *Are These Technologies the Future of the Constuction Industry.*
[19]Lavars, *How Drones Are Poised To Help Build the Cities of Tomorrow*
[20]Frey and Osborne, *The Future of Employment: How Susceptible Are Jobs To Computerization.*

CHAPTER THREE

EXECUTIVE FORESIGHT: A "HOW-TO" FOR CREATING THE FUTURE

Thinking about the future for some is seen as an exercise in futility due to the fact that the future cannot be fully known. This mindset must change in order to begin creating the future that will enable an organization to grow and thrive over the next 5 to 25 years. Success in Executive Foresight requires leaders and organizations to rethink what is possible and allow themselves and their team to entertain absurd notions about the future.

Impeding this process will limit their potential and prospects for dramatic growth. This is not to say that structure and parameters should be removed. Far from it. As you will see in the following pages, Executive Foresight is a structured process that creates highly focused narratives around critical questions that need answering.

One of the key objectives that many organizations have today is the ability to move from reactive posturing to proactive planning. Yet, the desire to do so and the skills to accomplish these objectives are sorely lacking. One of the main reasons for this is that old methods for strategic planning do not prepare organizations for the unknown. Old methods of strategic planning relied heavily on historical data in order to make projections into the future. This may have worked well 20 and 30 years ago when the rate of change was a fraction of what it is today. The future for the most part was fairly predictable. Additionally, unexpected events were often more of a manageable disturbance rather than a game changing disruption. Further complicating the planning process today is our unprecedented access to information.

Our decisions and outlook have always been determined by the following:

1. Access to information
2. Interpretation of that information
3. Integration of that information into a narrative about the future

Now, here is where the the past and present part ways...

In the past we had:

1. Limited or delayed access to information, which then led to a...
2. Narrower interpretation of that information, allowing for an...
3. Easier integration of that information into the cultural narrative ("good", "bad", "enemy", "ally")

Today we have:

1. Unprecedented and immediate access to information, which creates...
2. Multiple interpretations of that information, leading to...
3. A disjointed integration of that information into our cultural narrative ("good", "bad", "enemy", and "ally" all at the same time)

It is this same dilemma that leaders in business and organizations face today. In order to overcome this dilemma, executives must apply the principles of foresight to the decision-making process. This involves:

1. Leveraging the unprecedented access we have to information today.
2. Creating a contextual framework in order to better interpret that information.
3. Integrating that information into a powerful narrative that moves people to action.

SCANNING: LEVERAGING OUR UNPRECEDENTED ACCESS TO INFORMATION

In the Executive Foresight™ process, scanning is the means by which we leverage our unprecedented access to information. Scanning is a tool for finding specific points of data, information, or resources that help inform you of emerging trends, issues, or possible events. The scanning we did for the purposes of this book uncovered much of the information outlined in Chapters 1 and 2. The practice of scanning heightens your awareness of your environment and enables you to make sense of the massive amount of information you are expected to process everyday. It requires you to be intentional and think critically about this information. The more deliberate you are with your scanning efforts, the more it becomes a naturally occurring practice in your everyday life. We are constantly taking in new bits of information both consciously and subconsciously. Knowing what to do with that information is where most get hung up and the brain stores the new data away in a drawer labeled: "Irrelevant." With practice, you will begin to train your brain to make sense of seemingly unrelated points of information and translate them into a relevant and applicable narrative to guide your strategic planning process.

Jared's 3 Step Process For Scanning

Scanning can take on several forms, but to start out it is helpful to have a few formal guidelines to ensure that your efforts are maximized. The following are the most important areas to consider when scanning.

Step one: Identify an issue or critical question in order to establish the framework for your scanning efforts. Formulate this question around an existing trend, current issue, or actual event where the full implications are not yet known.

The following examples are the critical questions we asked for the purposes of this book:

"How will the recent growth in autonomous technologies impact the future of leadership and workforce development?"

"How will the utilization of Big Data impact the future of leadership and workforce development?"

Step two: Identify the "baseline assumption" answer to this question. We do this by posing and answering a question like:

"What do the majority of my colleagues and industry experts have to say about this?"

Step three: Find information that either confirms or contradicts the baseline assumptions being made or creates a new assumption altogether.

There are many practical tools and applications available for effectively cataloguing this information. However, the precise method of scanning is not as important as the adoption of scanning as a mindset. We are all inundated with information on a daily basis, being intentional about how you process that information is what makes the difference in today's world. The remainder of the Executive Foresight™ process focuses on how to use this information.

UNCOVERING IMPLICATIONS

Information gathered through scanning is still just information until you learn how to use it. The next step in the strategic foresight process provides you with a method for using the information you have gathered to create a contextual framework for the future. Creating a future context enables you to extrapolate the implications that specific trends and issues will have on the future as a whole. This framework will help you anticipate change and will better inform your actions today. However, it is important that this framework looks far beyond your organization, industry, and sector.

Too many organizations look for clues about the future within the bounds of industry forecasts and bias. When organizations solely rely on the internal industry expertise, they run headlong into the future with a false sense of security. Industry expertise is important, but the future of your industry and organization is dependent upon the future context within which it will operate. We live in an increasingly interconnected world, in which the implications of certain events are far more reaching than they once were. If we only look at how a certain technology, either one that is here today or one that is emerging, will impact the technology sector alone, then our perspective is far too narrow. In order to understand the real impact that something may have on the future as a whole, we have to explore how it will affect other facets of society.

A perfect example of this is the role that social media (specifically Twitter and Facebook) played in the Arab Spring, effectively overthrowing a 30+ year dictatorship in Egypt. Most wouldn't have guessed that these new social media tools could have had such far-reaching political implications.This is why when we consider how a possible trend or event might shape the future, we must establish a framework that includes all aspects of society. A useful tool for doing this is the classification system known as S.T.E.E.P., which looks at the Social, Technological, Economic, Environmental, and Political implications of a particular issue. Chapter 4 will demonstrate how we applied S.T.E.E.P. to our areas of focus.

DETERMINING IMPACT AND UNCERTAINTY

Once you have uncovered the potential implications of a specific trend, issue, or event, the next step is to determine which implications have the greatest potential impact (positive or negative) along with the greatest degree of uncertainty. Now this may seem counterintuitive, but you will want to focus on the greatest degree of uncertainty because such an issue or event is still far enough out in the future to take meaningful action today.

For instance, if an event has already begun to unfold or is on the brink of occurring, it is often too late to be proactive. The purpose of strategic foresight is to look far enough into the future that you have time to take meaningful action today. This means you must listen to "weak signals," and consider seemingly improbable potential events. These are often the areas in which you can have the greatest impact.

Too many organizations focus on high impact / high certainty events because they want to know how to best react to or recover from dramatic shifts and disruption. But if an organization truly wants to be more proactive in creating their future rather than reacting to it, then the leaders of these organizations need to reverse their thinking.

SCENARIO BUILDING: CREATING POWERFUL NARRATIVES

The implications with the greatest degree of impact and uncertainty are used to formulate the questions that will provide the basis for each future scenario that you create. Again, the point here is to look at situations that have not occurred, but have the potential to drastically alter the business landscape. You want to focus on implications that give you "lead-time" to make the best possible decisions for success in multiple future outcomes.

There are different approaches to creating good questions, but a simple way to do this is by using the "Will / Or" method, i.e., "Will this happen? Or, will that happen?" Each implication you select provides you with two alternative scenarios to build.

The purpose of scenario building is to identify key elements that will make each scenario believable and engaging. The last thing you want to create is a boring list of "possibilities" that are absent of imagination. One of the most important aspects of this process is to engage and drive creative thinking within your organization, so that you can apply the insights gained through this process to your present-day strategy.

This is not about predicting the future. Scenario building and planning have been used both by major corporations such as Royal Dutch Shell and by the US military in order to make better, more informed decisions in the present. Having the ability to put all the information that you uncovered into a plausible and relevant narrative for the future of your industry and organization helps to ensure your adaptability to sudden shifts in the market, and your resilience in the face of unexpected events. The end result gives you a significant competitive advantage.

LEVERAGING THE FUTURE

So often we make recommendations for change, strategic initiatives, or tactical action, but fail to present a convincing argument to support our recommendations. By completing this process from scanning to scenario building, you now have the format to overcome that obstacle. You now have the ability to see the future as an array of possibilities, of which you can take an active role in shaping.

The final step is to measure your current strategic priorities against each of the alternative scenarios you have created. The key here is to determine which scenarios your proposed strategies are most compatible with and to determine what actions need to be taken against those scenarios for which your strategies are incompatible.

Having given you a brief preview of how the Executive Foresight™ process works, in the next section you will discover how to put Executive Foresight™ into action.

CHAPTER FOUR

CREATING A CONTEXTUAL FRAMEWORK: IMPLICATIONS OF A.I., BIG DATA, AND AUTONOMOUS TECHNOLOGY

As we discussed in the previous chapter, in order to understand the implications that a trend, issue, or event may have on the future and, subsequently, on ourselves and our organizations, we must first uncover how those trends, issues, and events impact multiple facets of our society. The way we do this is through the S.T.E.E.P. categories.

Here is a brief description of each category and what it encompasses:

Social - This category looks at how an issue or event might impact social interaction, the way we relate to each other as people, and the opinions and inclinations that the public shares. We must look at how it affects the family, the community, and how people spend their time together.

Technological - Because most people consider the future or foresight as centering on new technologies, the "technology" category may seem a bit redundant. Despite technology becoming increasingly integrated into every facet of our lives, the advancement or growth of the technology sector is still very much vulnerable to outside trends, issues, and events.

Economic - When discussing the economic impact of a certain trend, issue, and event, we must think about specific areas where market confidence, financial stability, and access to capital are paramount to long-term growth and viability. We must also consider the impact on labor, currency, and education and training.

Environmental - Of all the different categories, environmental is the most important, because without the environment none of the other categories would actually exist. It is the category upon which all other categories depend; therefore, its impacts and implications are inherently far-reaching. For example, extreme weather events have a direct impact on the economy, our society, and the type of technology employed to respond.

Political – The political category centers on elected officials and governmental bodies that create and implement public policy. Key areas of focus in this realm include regulations, taxation, governance models, representation, and community engagement.

When using STEEP to explore the future implications of emerging trends and issues, it is important to consider a wide variety of alternative impacts. Do not be afraid to entertain seemingly absurd possibilities. As recent history has shown us, the future often does not unfold as we expect. The more creative and curious you are in this process, the more likely you are to capture the true range of potential outcomes.

A note about uncovering implications: You will notice that some of the potential implications contradict each other. This is neccessary in order to explore the full range of possibility. Also, certain implications may not seem relevant to the area of interest. As we've already discussed, implications in one area are often far more reaching than are immediately evident, e.g., technological advances have social implications, environmental impacts have political implications, etc. Additionally, some potential implications may not mesh with your view of the world. It is important that you allow your biases and assumptions to be challenged during this process. As you read the following implications, take note of your reactions and assess whether or not certain preconceived notions are inhibiting you from seeing a broader range of possibilities.

In order to uncover a wide variety of Big Data and Autonomous Technology implications, we answered a series of questions from each of the STEEP categories.

SOCIAL IMPLICATIONS

How might this impact our traditional concepts of the family (social significance, roles within, etc.)?

- Harmony in relationships may increase due to data driven compatibility algorithms, resulting in a significant decrease in divorce rates and single parent households.

- Alternative family structures such as polyamory, blended families, etc. become more prevalent as individuals rely more on computerized data analysis instead of social norms. As a result, the idea of traditional marriage and the traditional family structure will further diminish.
- Advancements in artificial intelligence and self-learning machines create a more intimate relationship between individuals and their devices, vehicles, homes, etc., shifting from utilitarian dependency to actual affection.
- Devices become so smart, opinionated, and well-informed that individuals begin to rely on them the same way they do family and feel about them the way they would feel about a beloved pet.
- Just as files can be transferred between devices via the cloud, so also would A.I. move among physical devices, omnipresent, adapting to its surroundings, learning from experience, and "maturing" alongside human beings. The human bond would then be with the A.I. and not the physical devices.

How might this impact the arts and entertainment and our outlets for leisurely activities?

- Leisurely activities abound in the age of information and interconnectedness, but not all leisurely activities are beneficial to all people. Too many options make it difficult to decide what will bring individuals fulfillment as well as long-term positive life learning experiences. Big data and analysis eliminates this obstacle.

- A.I.'s ability to process aggregate data about an individual's habits, environment, genetic history, susceptibility to illness or physical injury, and mental health issues provides unprecedented insights into what activities and pursuits would bring that person the greatest holistic fulfillment.
- Our devices, homes, vehicles, will all include an entertainment and leisure management system to help optimize downtime for maximum pleasure, stress relief, and personal fulfillment.

How might this redefine/magnify/deteriorate our traditional concepts of community both locally and nationally?

- Data driven compatibility for communities gives rise to a new concept for community and urban planning called "socio-datanomics," used to promote harmony and success within communities, based on common interests and complimentary assets and abilities.
- Instead of the socioeconomic factors that tend to determine the make-up of communities today, "socio-datanomics" would build communities around aggregate data that can simultaneously put the best interest of the individual and the best interest of society on the same track.
- Conversely, individuals and communities may choose to reject this type of model, based on its potential for discrimination (no one would really be able to know exactly what data was used to plan a community) or civil liberties violations (not having control over

privacy and how their personal data gets used).

- Alternatively, through data analysis, smart technologies are simply relied upon to determine where a person should live. Unlike the quizzes and assessments that already exist, greater trust is placed in the results due to advancements in A.I. and data analysis.
- Combined with the increased ability to work virtually, people are more likely to choose a community based on aggregate data.
- In the same way, cities and communities use aggregate data to attract and recruit specific individuals as residents.

TECHNOLOGICAL IMPLICATIONS

How might this impact the way we consume media and report information?

- Smart data identifies the type of media we need as an alternative to what we may impulsively want, e.g. the utilization of biometrics to improve your mood, stimulate creative thinking, reflection, compassion, etc.

How might this impact our means of transportation and physical mobility?

- The age of hyper-connectivity and "smart data" has truly arrived. Communication takes on a whole new definition through wearables and imbedded technology feeding a constant stream of information that enables our devices, vehicles, buildings, appliances, etc. to communicate with each other without our specific directive.

- The same goes for our infrastructure. "Smart roads," signs, and utilities rely on "smart data" to achieve a high level of communication between all connected devices (including ourselves), creating greater efficiency in transportation and logistics.

ECONOMIC IMPLICATIONS

How might this impact discretionary income, private investment, and consumer spending?

- Human-centric aggregate data creates far fewer and far better options to streamline consumer's decision-making process. This increases discretionary income as more people are making better decisions about where they work, where they live, what they buy, and where they vacation.
- Consumer spending is more predictable rather than sporadic. Data driven customization eliminates guesswork on the part of companies, thereby eliminating irrelevant advertisements to the consumer.
- Private investment increases as the risks associated with unknowns is greatly diminished. ROI, sustainability, long-term relevance, and social impact are all factored in with an extraordinary degree of accuracy.

How might this impact the demand for human labor, specialized skill sets, and talent?

- Human physical labor is replaced by human intellectual capital. There is no need for human labor when machines can operate themselves, fix themselves, and communicate with each other.
- Specialized skills and talents will revolve primarily around critical thinking, creativity, and new ideas. Virtually all quantitative and transactional tasks will be automated. Roles such as accountant, investment analyst, and insurance underwriter will all be unnecessary.
- The age of the service technician is over as machines are capable of learning from their mistakes and fixing themselves. Machines may still require human assistance from time to time, but will no longer depend on human intervention for overall maintenance.
- Big data analysis (rather than arbitrary opinions) will inform organizations of the type of cultural environment needed within the organization to achieve longterm success.
- Big data will increasingly enable companies to select the optimal employees for their organizational culture and goals.
- Due to increased automation of jobs and lack of employment options, a national basic income is implemented in lieu of expanding welfare programs.

How might this impact the demand for human labor, specialized skill sets, and talent?

- Human physical labor is replaced by human intellectual capital. There is no need for human labor when machines can operate themselves, fix themselves, and communicate with each other.
- Specialized skills and talents will revolve primarily around critical thinking, creativity, and new ideas. Virtually all quantitative and transactional tasks will be automated. Roles such as accountant, investment analyst, and insurance underwriter will all be unnecessary.
- The age of the service technician is over as machines are capable of learning from their mistakes and fixing themselves. Machines may still require human assistance from time to time, but will no longer depend on human intervention for overall maintenance.
- Big data analysis (rather than arbitrary opinions) will inform organizations of the type of cultural environment needed within the organization to achieve longterm success.
- Big data will increasingly enable companies to select the optimal employees for their organizational culture and goals.
- Due to increased automation of jobs and lack of employment options, a national basic income is implemented in lieu of expanding welfare programs.

How might this impact our access to, the quality of, and the perceived value of education?

- Problem solving, creativity, and critical thinking become more valuable as a skill set to organizations rather than the former emphasis on STEM skills. A.I. and robotics replace the need for STEM skills in human beings.
- Education acts as a conduit to nurture and preserve our humanity amidst a tech heavy quantified environment.
- A "quality" education is qualitative. Not focusing on how to quantify, organize, and analyze (tasks that are all easily computerized), but on fostering social and creative intelligence.

How might this impact our currency on a national and global level?

- Speculation no longer exists. Big data and A.I. allow for realtime and near accurate projections (along with a series of alternatives).
- Access to capital is globalized. Institutions and individuals can quickly determine the probability of default and repayment based on aggregate data and predictive analytics, utilizing information that goes far beyond one's credit score, such as your social circles, community, probability of influence both negative and positive, family connections and their net worth, health status, social status, social circles of influence, etc. All of this creates a highly detailed map, allowing institutions and individuals to make the best decisions.
- Currency beyond the money: social clout and contribution in the

form of art, service, innovation, etc., is "banked" and counted as currency to be used for the acquisition of goods and services, special privileges, etc. This also creates a culture of personal responsibility and incentivizes people to behave in a way that contributes to, rather than takes away from, society (you cannot leave your social clout in an estate).

ENVIRONMENTAL IMPLICATIONS

What impact might this have on the quality of our natural resources?

- Degradation of natural resources is prevented through systems that monitor realtime changes and communicate that information with each other in order to anticipate and prevent deterioration.
- Autonomous vehicles such as drones, become environmental guardians, monitoring the impact from the activities of companies and townships, reporting pollution, illegal dumping, excess use of water, and other valuable resources.

How might this impact the quantity of resources available to us today and in the future?

- Smart devices are used for monitoring realtime utilization of resources measured against aggregate data from multiple sources, e.g., meteorological, seismic, etc.
- Wearable and imbedded technology help eliminate waste and promote the responsible use of available resources. For example, my device monitors exactly what I need to consume in terms of

calories, water, nutrients, vitamins, etc., allowing us to healthily and sustainably live within the limits of our natural resources.

- Much of what we need will be delivered through autonomous smart vehicles and aerial transport drones. The A.I. will be developed to "know" what we need before we do, reducing the amount of time, energy, and fuel wasted by consumers in the shopping process.

How might this impact the quality of our food, distribution systems, farming, and the agricultural industry as a whole?

- Big data employs aggregate data from multiple channels, such as meteorological predictions and soil quality analysis, in conjunction with several other points of critical information, to help farmers and big agriculture determine which crops should be planted, when they should be planted, and where.
- Smart labels communicate with wearable and imbedded tech to determine the best options for the consumer, rating the product's caloric and nutrient intake in conjunction with that individual's metabolic and overall health and lifestyle data.
- Your device will also inform you of alternative options that may be better suited to your overall health and personal dietary preferences.
- Predictive analytics improves food distribution systems, creating highly accurate orders based on the meta-data of populations served. Overstock and waste is greatly eliminated, as consumer spending and demographic trends can be analyzed with a greater degree of accuracy.
- Autonomous farm equipment and smart data reduce mistakes

and dramatically increase yield, knowing when to harvest and at what time to get the optimal yield of each crop.

What impact might this have on the preservation/protection/destruction of wildlife and biodiversity?

- Smart monitoring and data analysis will help us better predict the environmental impact that the loss (or introduction) of certain plant or animal species will have on an area's ecosystem.
- The controlled elimination or introduction of a certain plant or animal species may also be employed to combat the effects of climate change, the dislocation and endangerment of species, or the loss of biodiversity.
- Drones and other forms of autonomous tech will operate as aerial shepherds, monitoring and preventing poaching, illegal dumping, overfishing, destructive farming practices, etc.

POLITICAL IMPLICATIONS

What regulatory impact might this have on business and community growth and development?

- The breadth of data will be wide and freely available, but the depth of data will be harder to access due to high costs or legal restrictions, such as intellectual property rights.
- Business development will depend on the depth of data available and the holding of licenses, certifications, etc. in order to stay within compliance. Licenses and certifications are necessary as

privacy will be the "litigation playground" of the future.

- A rapid acceleration of inequality resulting from unequal access to Big Data and analytics will be halted by regulation to democratize Big Data.
- Early adopters will excel where laggards will fall farther behind. Communities and businesses will be forced to adopt smart technology in order to survive the data driven hyperconnected future.
- Infrastructure will no longer support users not equipped with smart technology.

How might this impact taxation and revenue generation, both at the local and national levels?

- Government incentives for the early adoption of robotics and autonomous technology drives local economic development.
- Increase in local and national revenue due to the sharp increase in licenses needed for owning and operating robotic laborers and autonomous fleets.
- Introduction of the "robotics utilization tax" to fund the continued expansion and maintenance of a smart infrastructure.

All of the above implications amount to a large encompassing framework of the possible outcomes of Big Data and Autonomous Technologies. The next step in the process narrows the focus by determining which implications should be of greatest concern.

CHAPTER FIVE

ZEROING IN: GAUGING IMPACT AND UNCERTAINTY

The overall purpose of this exercise is to help establish the most pressing issues as it relates to the long-term success of the organization. The questions we answered in each of the STEEP categories identified potential implications and areas of uncertainty. Rather than listing each individual implication, we created simple statements that summed up the implications under each category. The next step was to decide which implications were of greatest concern based on two criteria: impact (the degree to which the implication would affect you or your organization) and uncertainty (the degree of uncertainty surrounding the implication).

We ranked each on a scale of 1 to 3.

1 = Little to no impact / uncertainty
2 = Some impact / uncertainty
3 = High impact / uncertainty

The chart on the next page illustrates our conclusions as to the impact and uncertainty of the STEEP implications we uncovered.

GAUGING IMPACT AND UNCERTAINTY

SOCIAL IMPLICATIONS	IMPACT	UNCERTAINTY
Harmony in relationships increased by data driven compatibility	1 ② 3	1 2 ③
Fulfilling leisure - greater insight into what will bring us the greatest holistic fulfillment.	1 ② 3	1 ② 3
Communities are developed in physical locations around "datanomics" using predictive planning to create ideal living and working spaces for the community - "Feng Shui 2.0"	1 2 ③	1 ② 3

TECHNOLOGICAL IMPLICATIONS	IMPACT	UNCERTAINTY
Our devices, vehicles, buildings, appliances, etc. are communicating with each other without our specific directive through the employment of "smart data," creating a high level of communication with each other that is predictive and ensures cooperation between vehicles and other means of transportation.	1 ② 3	1 ② 3

ECONOMIC IMPLICATIONS	IMPACT	UNCERTAINTY
Human physical labor is replaced by human intellectual capital, i.e. no need for human labor when machines can operate themselves, fix themselves, and communicate with each other.	1 2 ③	1 2 ③
Specialized skills and talents will revolve primarily around critical thinking, creativity, and new ideas. Input tasks and data analysis will be automated, such as bookkeeping, investment analysts, and basically any quantitative work.	1 2 ③	① 2 3
Big data will increasingly enable companies to select the optimal employees for their organizational culture and goals.	1 2 ③	① 2 3
Speculation no longer exists. Big data and A.I. allow for realtime and near accurate projections (along with a series of alternatives).	1 ② 3	1 ② 3

ENVIRONMENTAL IMPLICATIONS	IMPACT	UNCERTAINTY
Quantity goes up in conjunction with wearable and imbedded tech, e.g. my device monitors exactly what I need to take in such as calories, water, nutrients, vitamins, etc., allowing us to happily and healthily live within our means and needs, thereby eliminating waste.	1 (2) 3	1 (2) 3
Food quality (pre-packaged) - smart labels communicate with wearable and imbedded tech to determine the best options for the individual showing calories in conjunction with that person's metabolic data, lifestyle, etc., along with alternative options that may be better suited for their overall health and wellbeing.	1 (2) 3	1 2 (3)
Food distribution - significant decrease in wasted food / overstock. Predictive analytics creates highly accurate orders using meta-data on populations served, analyzing individual spending habits to form near-accurate estimations. This will also include demographic trends, such as population loss and growth.	1 (2) 3	1 2 (3)

POLITICAL IMPLICATIONS	IMPACT	UNCERTAINTY
Business growth and development - "Info is free, but knowledge is power" - the flow of data will be wide but the depth of data will come at a high price, both in access to it and the rights to use it.	1 (2) 3	1 (2) 3
Early adopters will excel where laggards will fall farther behind. Communities and businesses will be forced to adopt smart technology in order to survive the data driven hyperconnected future.	1 (2) 3	1 (2) 3
Government incentives for the early adoption of robotics and autonomous technology drives local economic development.	1 (2) 3	1 2 (3)

Having completed this exercise, the next step is to select the implication statements that are high impact / high uncertainty. Again, you want to focus on these areas of high uncertainty, as they provide the greatest degree of opportunity for being proactive. Based on our rankings of the above implications, we selected Social Implication #3 and Economic Implication #1 as our two most important areas of focus. Having established our areas of focus, we were ready to begin the scenario-building process.

CHAPTER SIX

SCENARIO BUILDING

There are several ways to write a scenario. The method an organization chooses is heavily dependent on its objectives for the project at hand. The following is a simplified version of the process we used for creating our scenarios.

CREATING YOUR WILL/OR QUESTIONS

The scenario building process begins with formulating "Will / Or" questions, using the implications statements you found to be of greatest importance. Each statement becomes a "Will" question. This "Will" question must then be challenged with an "Or" question, which will require further creative thinking on the part of the scenario building team. The implications statements we selected provided the basis for the following "Will / Or" questions:

Economic Implications #1

1. Will the employment of autonomous intelligent machines (machines that can operate themselves, fix themselves, and communicate with each other) redefine the value of human contribution to the workforce from skilled physical labor to human intellectual capital?
2. Or, will the growth and adoption of autonomous technologies be slowed by public concern about the safety and benefits of artificial intelligence?

Social Implications #3

1. Will the continued trends toward sustainable development and mass customization through the utilization of big data completely transform the way we design, build, site select, and allocate resources for the development of physical communities?
2. Or, will smart community planning be stalled by public concern over privacy and the potential for discrimination?

The purpose of scenario building is to identify key elements that will make each scenario believable and engaging. As stated in chapter 3, the last thing you want to create is a boring list of "possibilities" that are absent of imagination. In order to give your scenarios "life," place yourself 10 to 25 years into the future and imagine you are a historian recounting the events surrounding this phenomenon. We used the following questions to help build context for our scenarios as well as to identify the obstacles and opportunities each scenario presents.

JARED'S FIVE STEP PROCESS FOR SCENARIO BUILDING

Step one: What happened and when? - Start by choosing a timeframe that sets the stage for your scenario and fits with your chosen title.

Step two: How did it happen? Why? – What STEEP elements were present beforehand that played a key part in bringing this scenario to fruition? What indicators were present beforehand that, if addressed, could have prevented this scenario from taking place?

Step three: Who wins? Who loses? – Which segment of the population, age demographic, industry, sector, etc. came out ahead and which have suffered significant setbacks?

Step four: What did they do or not do? – What specific action did (or didn't) those who thrived in this scenario take to ensure their success?

Step five: Where are they now? – Where are both groups now? Are they thriving, struggling, or evolving? Why?

Remember that each of the scenarios you create should present a very different future outcome. Additionally, incorporating other trends that you uncovered through scanning into your narrative will make your scenario more effective. In doing so, you will have a much richer narrative, by which new points of information may emerge. This will give you greater insight into how to prepare for potential disruptors and transform them into new opportunities.

CHAPTER SEVEN

A SERIES OF ALTERNATIVES: FOUR FUTURES FOR THE 21ST CENTURY WORKFORCE

SCENARIO 1: CREATIVE DESTRUCTION

The employment of autonomous intelligent machines (machines that can operate themselves, fix themselves, and communicate with each other) redefines the value of human contribution to the workforce from skilled physical labor to human intellectual capital.

The VW emissions software scandal in 2015 raised the alarm to the hackability and manipulation of vehicle software in order to evade regulation and falsify data. This incident did very little to reassure consumers that corporations can be trusted to serve the interests of the public, the environment, and an overall commitment to excellence when profit is of primary concern. This concern extended into the public sector, when the Flint, MI water crisis exposed gross negligence on the part of local governments to provide its citizens with safe drinking water. This crisis was followed by several others in cities around the country. This served only to reinforce the growing distrust of the population, especially Millennials, in both public and private institutions.

What was most interesting to historians was the clear ideological shift that occurred, differentiating the upcoming Millennial generation from its predecessors. Unlike the Boomer generation who intrinsically distrusted technology, Millennials and younger Gen X'ers became increasingly distrustful of the people in charge, not the technology. As the Millennial voting block began to surpass that of older generations, they began to increasingly push for greater transparency in both corporate and government entities. Their hope was to use technology, specifically the development of smart technologies, to achieve this transparency. Under pressure from their constituents, Congress passed legislation to accelerate the development of an artificial intelligence that could be integrated into the nation's infrastructure in order to provide realtime data about the safety of vehicles, roadways, public transit, pipelines, and water systems.

"We aren't going to wait around for the corporations or local governments to take the initiative. No, we've seen what that looks like and the fact is that without federal government intervention, corporations and municipalities will undoubtedly make decisions that serve their bottom line even if it means putting consumers and citizens at risk." (from archived interview)

This led to the formation of the U.S. Department of Technology and Infrastructure. The DOTI's primary responsibility was to oversee the rebuilding of the U.S. infrastructure outfitted with software developed by the U.S. government cleverly titled "Babel." The purpose of Babel was to ensure that all devices, vehicles, structures, roadways, etc., spoke the same language. Any device, vehicle, roadway, or structure designed for public use that was not outfitted with Babel was not allowed to travel on or intersect with major highways and interstates as this would create public safety concerns. Individuals were essentially pushed into either retrofitting older model vehicles or upgrading. In order to ensure public safety, this smart technology was also integrated into the water systems of all U.S. cities through smart pipes that provided realtime analysis on the safety of drinking water.

Similar technologies spread through the private sector, accelerating the development of autonomous machines and further integrating artificial intelligence into our daily lives. This brought about a dramatic shift in the workplace, as more and more humans were replaced by computerization. Between 2015 and 2035 the U.S. saw 47% of jobs replaced by automation. Despite a clear need to dramatically overhaul the education system in order to equip the future workforce with skills that would meet the demands of a high tech future, political polarization and economic uncertainty stalled any meaningful advances in training and development.

This created a unique problem within the construction industry, as the call to rebuild the U.S. infrastructure created a massive demand for highly skilled labor. Unlike in previous years, the construction industry needed talent based on intellectual capital, rather than skilled physical labor. In a world where machines were capable of operating themselves, learning from successes and mistakes, and fixing themselves, operators and service technicians were no longer in demand. Skills and attributes such as creative thinking and problem-solving were necessary for contractors and developers to collaborate with their machines and stay ahead of the competition. As fewer workers were needed, the existing workforce needed to be skilled across disciplines and highly adaptable to a variety of tasks.

The rapid advancement of technology brought a significant strain to the small and mid-size market contractors as they were ill-equipped to compete with the larger companies when it came to major infrastructure projects. Contractors who adopted new autonomous technologies and learned early on how to capture, analyze, and utilize customer data were at a greater advantage when it came to acquiring contracts and new projects.

One of the greatest areas of opportunity for the industry was in the retrofitting of existing public and private structures, such as roadways, water systems, buildings, homes, and nearly anything that needed to be able to communicate with another object. It was here that the industry truly began to swell with new entrants in the small and mid-size contractor market. This created problems for the legacy contractors. In addition to the new competition, legacy contractors were still scrambling to catch up to where they needed to be in order to remain competitive. These companies had failed early on to make the necessary changes to their business practices as it related to talent and skills development. Instead, they found themselves scrambling for talent from other industries and sectors.

IMPACT ON WORKFORCE DEVELOPMENT: Fewer jobs resulted in high unemployment. Highly skilled workers were in demand with technological skills enabling them to understand the capabilities and limitations of computerized systems in order to complement these systems in the workplace. Workers needed cross-disciplinary skills. Fewer workers meant existing workers needed a broader skill set and the ability to operate at the intersection of different fields of expertise, e.g. science, technology, and sales. Critical analysis, creative thinking, and adaptability were all highly valued. While computers excel at analytical processes, due to the rigidity of A.I., humans were still needed to think, quite literally, outside the box.

SCENARIO 2: THE GREAT LOCKDOWN

The growth and adoption of autonomous technologies is slowed by public concern about the safety and benefits of artificial intelligence.

The VW emissions software scandal in 2015 raised the alarm to the hack-ability and manipulation of vehicle software in order to evade regulation and falsify data. This was the first in a string of corporate corruption scandals that revealed to the public how easy it had become through intelligent machines to manipulate data at the public's expense. As artificial intelligence began to infiltrate every area of consumer's lives, from vehicle and home operations to the types of goods and services offered to the monitoring of public and environmental safety, the capacity of corporate entities to use or manipulate data for financial gain had greatly surpassed the government's ability to regulate it.

Public outcry put the pressure on the U.S. government to establish greater oversight in the development of smart technology. The EPA's response to the scandals brought in greater regulations and scrutiny over the type of technology and software that was being integrated into our vehicles, homes, and other devices. The further development of smart technology was halted as the government implemented regulations to safeguard consumers and the public at large. The result of this put the promise of self-driving cars on hold as technology companies scrambled to work on making A.I. safer.

"The problem that arose from these scandals was that the public lost trust in the data that was being reported. If they could manipulate the data in my vehicle then what about the data in my home? Is my home monitoring system being hacked? Is my home really consuming the reported amount of energy?" (from archived interview)

Questions like these created enough doubt in the reliability and safety of smart technology that investors began to pull funding until they and the public could be assured that these vulnerabilities could be eliminated. Tech giants like Google, Amazon, Apple, and Microsoft saw this as an opportunity to lock up their operating systems so that their devices were only able to operate and communicate with each other. It was a flashback to the early years of Silicon Valley and the battle over proprietary software. The appetite for autonomous technology was still alive and well, but users had to essentially pick sides. The "great lockdown" in operating systems created problems with consumer devices such as smart phones and tablets.

"The failure to work together to make A.I. safe has caused a near reversal of the greatest technological advancements made up to this point in human history." (from archived interview)

This reversal however was not a loss for everyone. Where there is a crisis, there is opportunity and no one knew this better than the tech giants of Silicon Valley. Tech companies now had greater control over their systems which also meant greater control over the data collected on their users. The promise of security has created some of the greatest wealth in all of human history and this time it was no different. The concern over the vulnerability of vehicles and other devices was the driving force that created a new (old) type of autonomous machine. Rather than depending on GPS (an open platform technology), these machines learned to become spatially aware. At the time, this technology was nothing new. In fact, research done at MIT had made significant advancements in this field within the first decade of the 21st Century. But at that time their advancements were overshadowed by the exciting promise of a world that operated online and open source, allowing humans and devices to communicate automatically.

The benefits of automation had already been seen in several industries including construction, agriculture, and heavy equipment. Tech companies were able to partner with heavy equipment manufacturers like Caterpillar and Komatsu to expand autonomous fleets on a closed loop enterprise platform. These partnerships helped to make the transition to "high tech" much easier for small to mid-sized contractors as they were more comfortable with getting their full range of services from one provider. The concerns that many contractors had in the past about adopting the "latest and greatest" technology were alleviated as they no longer had to worry about cross-platform integration, updates, patches, and various other fixes that created more problems and slowed operational efficiency. Despite being beholden to one manufacturer's system, contractors found it was a reasonable price to pay to reduce the headache, inefficiencies, and risks involved in a cross- or open-platform system.

Fears that technology would completely replace the need for skilled physical labor were also put to rest as the skills of service technicians, engineers, and equipment operators were able to keep pace with the technological demands. Another benefit that came about as a result of this situation was that sales organizations were able to compete on their quality of service and strength of relationships rather than the features and benefits of some new technology. "Strong interpersonal connection is at the core of what makes us human. The software scandals that triggered public outcry and oversight were a blessing in disguise as it allowed us to reconnect with each other as people." (from archived interview) As such, organizations had to rethink the type of talent they needed in order to be successful.

IMPACT ON WORKFORCE DEVELOPMENT: Organizations sought out individuals who were emotionally intelligent, highly organized, excellent communicators, and creative problem solvers. Jobs that would have been relegated to computers, requiring computational skills and data analysis, continued to be filled by humans. Educational institutions were better able to keep pace with workforce demands, and organizations carried less of the training burden. Technological skills were still greatly valued as the various competitors strove to out-innovate each other. However, these workers were valued specifically for their computer and analytical skills and not needed to work across disciplines.

SCENARIO 3: THE DEMOCRATIZATION OF BIG DATA

The continued trends toward sustainable development and mass customization through the utilization of big data completely transform the way we design, build, site select, and allocate resources for the development of physical communities.

The big data revolution ushered in an entirely new way in which we viewed community development. The global surge in population in the early part of the 21st century created a significant strain on natural resources and allocation of those resources to various communities. However with corporate profits as the biggest catalyst of development and insufficient regulations, building and development remained anything but sustainable. If society was going to overcome these unprecedented environmental challenges, the public needed a larger stake in the planning and development decision-making process.

Essentially, the scales of power needed to be balanced through a leveling of the playing field. The public began to push for the democratization of Big Data as a means to achieving this. Making the information that fueled the success of so many large companies available to the public not only gave individuals greater power to make smarter decisions for themselves and their communities, it also unleashed a wave of innovation in the fields of data analytics and artificial intelligence.

This paved the way for the global development of the open source project for big data democratization and analysis known as "The Omniscience Project." The free flow of information and access to data allowed developers to advance the capabilities of artificial intelligence through carefully designed algorithms which could interpret, analyze, and present a series of options and ideal outcomes to key decisions made by individuals, corporations, communities, and governments. Before this advancement, access to data meant very little as the average individual or small organization did not possess the needed tools or expertise to understand and utilize Big Data. However, due to the advancements in artificial intelligence, an individual or community could now make meaningful decisions and accurately determine what options best served the public interest.

It goes without saying that many were skeptical at first and feared they would lose their identity, privacy, and choice when it came to how and where they lived their lives. However, as citizens allowed their personal data, (habits, genetics, etc.) to be integrated into The Omniscience

Project, developers were able to create and market communities that matched the optimal conditions that would allow for all people to truly embrace their right to life, liberty, and the pursuit of happiness. And that was the brilliance behind the program. The Omniscience Project was created to embody the principles upon which our democracy was based. This of course changed the focus of public discourse from a fear of losing one's privacy to the freedom to reach one's full potential.

In order to ensure that data remained free and open, a public-private partnership was developed to create an algorithm that put the interest of the public front and center and created oversight within corporations, municipalities, and any and all entities that served the public in order to preserve and reassure the trust that data was not being used for personal gain, malicious intent, or to disenfranchise certain segments of the population.

This was set in place (at least theoretically) to ensure that private interests could not make decisions that were not inline with the public's best interest. Every financial transaction was scrutinized by the system void of human input and intervention. No longer could companies or individuals "cook the books" or fudge the numbers when it came to impact on the public, the environment, or the financial system. Incidents such as the financial collapse of 2008 and the BP oil spill in 2010 were key examples used in the justification of creating and enforcing the algorithmic model of governance. The predictive power of The Omniscience Project prevented private companies from taking such reckless risks with the public's wellbeing.

One major benefit that came out of this era was the development of smart communities. This access to information by the public and their ability to now make smarter and better decisions paved the way for massive advancements in new construction and community development. With the tools made available through The Omniscience Project, construction companies and developers were now able to take most of the guesswork and decision-making out of the planning and building process. This created greater operational efficiency, reduced waste, more accurately predicted costs and profits, while aligning the needs and interests of the builders with their consumers. Additionally, the newly created structures were built to serve even the unrealized needs of the public not only today, but well into the future with as little impact as possible to the environment.

IMPACT ON WORKFORCE DEVELOPMENT: Organizations used the predictive power of data analytics to make better decisions in hiring; thereby, improving corporate culture, employee retention, and productivity. Employee satisfaction and engagement increased overall. Technological and analytical skills were greatly valued to help make sense of the large amounts of information available and to maintain and ensure the integrity of the system. With greater value placed on transparency, trust, and the public good, the accrual of personal wealth became less important. Employees were less motivated by compensation and instead looked for opportunities for personal fulfillment through both their work and personal lives.

SCENARIO 4: UNNATURAL SELECTION

Smart community planning is stalled by public concern over privacy and the potential for discrimination.

The Industrial Revolution ushered in unprecedented progress in human ingenuity and longevity. By the late 20th century, much of humanity had been enjoying the age of limitless growth and infinite possibility for over 200 years. The inevitable question that lurked in the minds of scholars and historians was, "Yes, but for how long?" Some began to suggest that the current systems of government and economic policy were not capable of recognizing or addressing the very real limits to growth imposed by finite resources.

The global surge in population throughout the latter 20th and early part of the 21st century created a significant strain on natural resources and allocation of those resources to various communities. As the wealthier nations of the world began to feel the effects of growing populations amidst rampant environmental degradation, even the basic elements of survival, such as water, began to seem at risk. It became clear that international actions needed to be taken in order to safeguard the future of humanity. In 2023, the United Nations Conference on Sustainable Development met to construct a proposal for overcoming these environmental challenges. Among the proposed solutions for ensuring that resource allocation and conservation efforts were both sustainable and fair, U.S. delegates, with support from corporate partners, presented a plan for smart community development called

The Omniscience Project. This system would have unfettered access to public data fed into a "socially and environmentally conscious" algorithm.

The plan was fairly straight forward. Private corporations would work together with the government to share data they held on their customers in order to pinpoint population needs and requirements as it pertained to natural resources such as water, farmable land, energy, etc. In order for this proposed plan to work the public would be required to release all rights to their personal data in order to help ensure a high degree of accuracy when it came to where resources, land allocation, new development, and investment in infrastructure should take place.

Despite the fact that the public had been giving away their information for free for over two decades, the reality that they would have to willingly release all of that information (including those data points that reveal things about them that they themselves were unaware of) created a great deal of concern about the potential risks and benefits of this arrangement.

One of the greatest voices of opposition to this proposed plan came from those who had already been working to rein in the unregulated flow of the public's personal information.

The Electronic Privacy Information Center (EPIC) raised concerns about the negative impact that predictive analytics will have on the individual's right to privacy as well as other freedoms. Despite the fact that overt discrimination had been illegal for several decades, many were concerned that the plan's proposed utilization of big data analytics would be used to automate discrimination. This concern was not unjustified. Past experiences with corporations that oversaw the financing of housing and development created great concern that this new project would only serve the more advantaged portion of the population.

"When we allow our spending habits, financial history, health records, lifestyle choices, friends and family, etc., to be analyzed by an algorithm we risk creating a reward system that is inherently unequal. Those who have been in a position of upward mobility and have the time and capital to better themselves will inevitably be seen as more worthy." (from archived interview)

As public opinion moved against the proposed Omniscience Project, private corporations pulled back and began to focus their time and energy on meeting the market demand for increased privacy and personal data security. Cybersecurity became the greatest concern at the start of the big data revolution, but the rapid advancement of new technologies and learning algorithms made it difficult for the U.S. government and tech companies to stay ahead of security breaches and espionage. Now, consumers opted for security over convenience in their day to day lives. The public wanted to feel that their personhood, both physical and digital, was safe from exploitation and manipulation. This was especially true when it came to their homes, families, and their ability to achieve personal advancement.

With a public solution off the table, private industries began to propose solutions for mitigating dwindling resources. Community development and construction continued down the path it had for many years, but with one key difference - tech companies were now in the construction business. Tech giants realized they could take the concepts of the proposed Omniscience Project and create a series of new agreements that enabled these companies to provide services, resources, and smarter community development plans that met both the current and future needs of their customers. The construction industry's slow-to-adopt mentality made it ripe for disruption. Now Facebook, Apple, Amazon, and Google expanded their ecosystems from devices and cloud-based services to physical communities designed around not just the individual user, but entire communities of users.

"The fact that construction companies and general contractors were historically late to adopt new technology made the decision easy. We needed to find a solution that would help ensure a sustainable and healthy future for the American people and that meant we needed to completely rethink the way we designed and built our communities. We already had the data and we knew that the idea of smart community living was of great interest to the public. We were the only sensible solution for quickly bringing these ideas into reality while maintaining a great deal of transparency with our customers. It truly has become a partnership..." (from archived interview)

IMPACT ON WORKFORCE DEVELOPMENT: As the tech industry disrupted the construction industry, a large premium was placed on workers who had engineering and design skills in addition to programming skills and tech savvy. Creativity was a big factor since tech companies needed to put their brand on the communities and structures they were designing. Data analysis and design innovation were also key, as the tech industry leveraged the information they had about consumers to create the kind of environments in which consumers wanted to live. Much like in the past, organizations needed competitive compensation and benefit plans along with attractive work environments. As early adopters in these areas, tech companies led the way.

CHAPTER EIGHT

LEVERAGING THE FUTURE

How do we leverage what we now know is possible about the future?

While we can't know what the future will bring, we can follow certain indicators to uncover a spectrum of potential futures, as we did with the creation of our alternative future scenarios. This process has little value though, if we don't use this information to evaluate our present day strategies. Where you currently invest your time, energy, and capital reveal your assumptions about the future, whether your realize it or not. But, what about the other potential futures, such as the ones we created in our scenarios? In order to increase their viability, your strategic investments must be measured against multiple possible futures. What assumptions do your current strategic priorities make about the future? How will these strategies play out in the other potential futures we have created? The following table is a good tool for analyzing your strategies across the board, from capital and resource investment to company culture and values. It will enable you to identify your strengths and weaknesses in each strategic area across the four potential future scenarios.

The table below gives a general list of strategic priorities in the vertical column; however, this list can be customized to be more relevant to the specific strategies of an organization. You can then work through each scenario and strategy to determine how compatible the two are. Use the following color code to rate each strategy:

Green - Strategy is very compatible with scenario

Yellow - Strategy is mostly compatible with scenario

Red - Strategy is not compatible with scenario

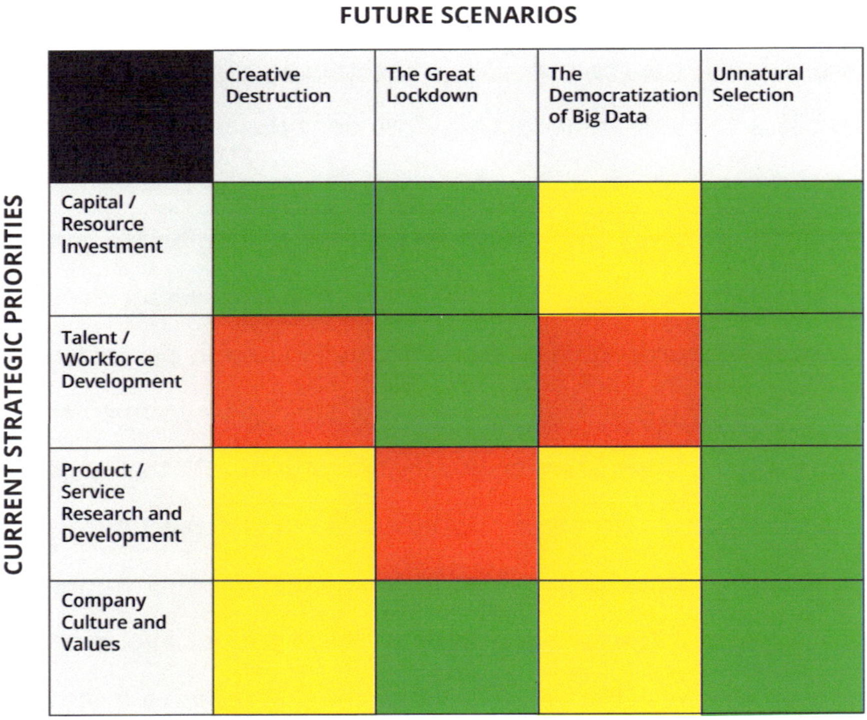

For strategies scoring in the red, you should ask yourself the following questions:

1. What indicators should we watch out for that would signal this future scenario's likelihood?
2. What should we do today to ensure that we are prepared for this future scenario or one that is very similar to it?

For strategies scoring in the yellow, you should consider these questions:

1. Where do we need to focus our energy in order to ensure our future growth and long-term success?
2. What potential surprises lurking in this scenario should we be aware of?

STRENGTHENING YOUR STRATEGIC PRIORITIES

Having rated your strategic priorities against multiple possible futures, you will probably become aware of potential deficiencies. The following considerations in the four strategic areas listed above (Capital / Resource Investment, Talent and Workforce Development, Product / Service Research and Development, and Company Culture and Values) will help you to begin the process of strengthening your organization's strategies.

Capital / Resource Investment

Where we spend money is a clear indication of priorities. Regardless of what we might think is priority, the reality is that every investment we make is based on an assumption about the future. Consider how your investments reflect your beliefs about he future by answering the following questions:

1. Do you believe that very little will change in the next 5–10 years or are your investments shifting and adapting to anticipated changes?
2. Are you anticipating new players that could significantly disrupt your business or the industry as a whole? How do you plan to leverage your resources to avoid or counteract this possibility?

It's questions like these that must be answered before you can truly determine how strong or weak your growth strategies are in multiple future outcomes.

Talent and Workforce Development

What type of talent do you need to attract in order to thrive in multiple future outcomes? The best place to start answering this question is by looking at the type of talent you are attracting right now. For some organizations, talent attraction is more serendipitous than strategic. But for others, attracting top talent is the greatest area of concern as they look to the future. Many organizations focus on competitive compensation and rich benefit plans while others focus on professional development, corporate culture, and flexible work hours. Regardless of which approach you are taking, your strategic priorities in this area will determine the type of candidates you are attracting and the type of talent you are retaining. Consider the following:

1. What transferrable skills and expertise do we have in our organization today, i.e. those which are not limited to our market or industry?
2. Is there untapped intellectual capital within our organization?
3. How do you plan to expand the capabilities of your workforce to meet shifting demands?

These questions will help you identify potential skills gaps as you project your workforce development strategy into various possible futures.

Product / Service Research and Development

"If it ain't broke...break it." The reality is that if you don't break it, someone else will. Ask yourself the following questions:

1. How confident am I that our products and services will still be relevant in 5, 10, or 20 years?
2. What factors distinguish our products and services from our competitors?
3. Do our products address a real need or are they a nicety?

Answering these questions will give you a much greater picture of where your strategic priorities are today and where they should be tomorrow.

Company Culture and Values

Many organizations put a high premium on their stated values and company culture. This can be used to attract talent, ensure customer loyalty, stimulate creativity, encourage productivity, and promote leadership, among other things. However, an overbearing or rigid culture can make it very difficult to adapt to rapid and unexpected changes. Regardless of the type of culture your organization has created, the following questions should be considered:

1. How do our company values align with the needs and issues in today's business environment?
2. How do our company values and/or cultural characteristics translate into actionable strategies that our customers can use to ensure their long-term success?

These questions must be asked on a regular basis to clarify company values and meaningfully engage employees and customers. Only having done this, can you look at how these cultural values play out across potential future outcomes.

START CREATING YOUR FUTURE TODAY

Once you've completed this process, you will be able to share with the members of your organization a clearer vision of the obstacles and opportunities ahead, ensuring their greater understanding of and investment in the direction of the organization. You will also be ready to use what you've learned to create a culture of foresight and innovation within your organization. This is important because Executive Foresight should never be considered a one-off process, but a new way of looking at the world. You will use these tools again and again to stay ahead of changes and create the future you want to see. As such, you will begin to build the kind of organization that employees want to work for and competitors strive to become.

Regardless of how uncertain, volatile, and unstable the 21st century may be, you have the power to create continuous success in your business and community. This will require you to be vigilant about what is on the horizon, scrutinize the data you use to make long-term strategic decisions, and be willing to take the necessary steps to ensure your work's success.

REFERENCES

1. "Confidence in Institutions," Gallup. See http://www.gallup.com/poll/1597/confidence-institutions.aspx.
2. Brandon Busteed, "America's 'No Confidence" Vote on College Grad's Work Readiness," Gallup (April 24, 2015).
3. "Workforce 2020: The Looming Talent Crisis," SAP and Oxford Economics. See http://www.successfactors.com/en_us/lp/workforce-2020-insights.html.
4. "Life Sciences Workforce Trends Report 2014," The Coalition of State Bioscience Institutes. See http://media.wix.com/ugd/dd6885_ce2f885617d147e5a1e942979298c0fd.pdf.
5. "The Millennial Leadership Survery," WorkplaceTrend (July 20, 2015). See https://workplacetrends.com/the-millennial-leadership-survey/.
6. Ibid.
7. Dan Schawbel, "10 Workplace Trends You'll See in 2016," Forbes (November 1, 2015). See http://www.forbes.com/sites/danschawbel/2015/11/01/10-workplace-trends-for-2016/#1619cc37222c.
8. Ibid.
9. Nick Petrie, "Future Trends in Leadership Development," Center for Creative Leadership (accessed July 28, 2015). See http://insights.ccl.org/wp-content/uploads/2015/04/futureTrends.pdf.

10. Neil Irwin, "With 'Gigs' Instead of Jobs, Workers Bear New Burdens," NY Times (March 31, 2016). See http://www.nytimes.com/2016/03/31/upshot/contractors-and-temps-accounted-for-all-of-the-growth-in-employment-in-the-last-decade.html.
11. Ibid.
12. Ibid.
13. David Barista, "The Big Data Revolution: How Data-Driven Design Is Transforming Project Planning," Building, Design, and Construction (February 14, 2014). See http://www.bdcnetwork.com/big-data-revolution-how-data-driven-design-transforming-project-planning.
14. Ellen Huet, "Server and Protect: Predictive Policing Firm PredPol Promises To Map Crime Before It Happens," Forbes (March 2, 2015). See http://www.forbes.com/sites/ellenhuet/2015/02/11/predpol-predictive-policing/?linkId=12286273#1a2763b3407f.
15. Miroslaw Skibniewski and Mani Golparvar-Fard, "Toward a Science of Autonomy for Physical Systems: Construction," Computing Community Consortium (June 25, 2015). See http://cra.org/ccc/wp-content/uploads/sites/2/2015/07/Construction-v4.pdf.
16. "30 Corporations Working on Autonomous Vehicles," CB Insights. (Accessed June 4, 2016). See https://www.cbinsights.com/blog/autonomous-driverless-vehicles-corporations-list/.
17. Ibid.
18. Kimberly Hegemony, "Are These Technologies the Future of the Construction Industry?" ForConstructionPros.com. (May 14, 2015). See http://www.forconstructionpros.com/blog/12074300/are-these-

technologies-the-future-of-the-construction-industry.

19. Nick Lavars, "How Drones Are Poised to Help Build the Cities of Tomorrow," Gizmag. (March 1, 2015). See http://www.gizmag.com/drones-building-construction-industry/36306/.
20. Carl Benedikt Frey and Michael A. Osborne, "The Future of Employment: How Susceptible Are Jobs to Computerization?" University of Oxford (September 17, 2013). See http://www.oxfordmartin.ox.ac.uk/downloads/academic/The_Future_of_Employment.pdf.

APPENDIX: SCAN HITS

7 Workforce Management Trends That Will Impact IT in 2015: http://www.cio.com/article/2881184/careers-staffing/7-workforce-management-trends-that-will-impact-it-in-2015.html

Workforce on demand: Are you ready?: http://dupress.com/articles/on-demand-workforce-human-capital-trends-2015/

Weekly Wrap: 6 Critical Workforce Issues Companies Face: http://www.eremedia.com/tlnt/weekly-wrap-6-critical-workforce-issues-companies-face/

Cramer: What United Rentals Teaches Us: http://realmoney.thestreet.com/articles/01/23/2015/cramer-what-united-rentals-teaches-us

10 Workplace Trends For 2015: http://www.forbes.com/sites/danschawbel/2014/10/29/the-top-10-workplace-trends-for-2015/#4d34566b654b

Make Use of Construction Equipment Trends to Maximize Profit: http://www.forconstructionpros.com/article/12063299/make-use-of-construction-equipment-trends-to-maximize-profit

The Big Data Advantage for Construction | Backing America's Backbone: http://americasbackbone.com/plan/the-big-data-advantage-for-construction/

Secrets of Big Data: http://www.agriculture.com/news/business/secrets-of-big-data_5-ar49267

Obama's new overtime rules: How they'd work and who they'd affect: http://www.latimes.com/business/la-fi-obama-overtime-rules-explainer-20150630-htmlstory.html

Equipment Makers De-Feature for Rising Rental Market | ENR: http://www.enr.com/articles/9634-equipment-makers-de-feature-for-rising-rental-market?v=preview

How drones are poised to help build the cities of tomorrow: http://www.gizmag.com/drones-building-construction-industry/36306/

Construction Technology: From the Cloud to A.I.: http://www.zlien.com/articles/construction-technology-cloud/

Technology Trends Converge to Transform the Construction Industry: http://enewsletters.constructionexec.com/techtrends/2013/09/technology-trends-converge-to-transform-the-construction-industry/
Trends Impacting Pharma Equipment: http://www.pharmamanufacturing.com/articles/2015/trends-impacting-pharma-equipment/

Life Sciences Industry Forecast: http://www2.deloitte.com/content/dam/Deloitte/global/Documents/Life-Sciences-Health-Care/gx-lshc-2015-life-sciences-report.pdf

YC-Backed EquipmentShare Is Like Airbnb For Construction Equipment: http://techcrunch.com/2015/03/09/equipmentshare/

2015 Equipment Rental Industry Forecast to Grow 8.1%: http://www.marketwatch.com/story/2015-equipment-rental-industry-forecast-to-grow-81-2015-01-19

There's No New Way for the Construction Market to Move Dirt — Or Is There?: http://www.forconstructionpros.com/article/12020677/theres-no-new-way-for-the-construction-market-to-move-dirt-or-is-there

Economic Recovery Leads to More Equipment Rentals: http://enewsletters.constructionexec.com/managingyourbusiness/2014/11/economic-recovery-leads-to-more-equipment-rentals/

The Top Six Construction Trends for 2015: http://blog.capterra.com/top-six-construction-trends-2015/

Heavy Equipment Trends: Rentals Are Up, New Equipment Is Preferred and Adaptability Is Key: http://enewsletters.constructionexec.com/managingyourbusiness/2014/11/heavy-equipment-trends-rentals-are-up-new-equipment-is-preferred-and-adaptability-is-key/

State of the Industry: Rentals to Reach $40 Billion in 2015: http://www.forconstructionpros.com/article/12024265/state-of-the-industry-rentals-to-reach-40-billion-in-2015

The Power and Potential of Telematics | Construction Equipment: http://www.constructionequipment.com/power-and-potential-telematics

Rent Your Robotics and Other Construction Equipment from BuildingPoint Mid-America: http://bpmidamerica.com/general-contractor-solutions-and-services/buildingpoint-construction-equipment-rental/

7 Trends Driving Global Health And Life Sciences In 2015: http://www.forbes.com/sites/nicolefisher/2014/11/21/7-trends-driving-global-health-and-life-science-in-2015/#2030d1d22cc5

7 Technology Trends That Will Impact The Life Sciences in 2015: http://blog.montrium.com/blog/7-technology-trends-that-will-impact-the-life-sciences-in-2015

Gartner on IT Trends in Pharma - Bio-IT World: http://www.bio-itworld.com/BioIT_Article.aspx?id=27172

CSBI Life sciences workforce trends: http://www.csbioinstitutes.org/

From Diversity to Dogs: The Top 10 workplace trends for 2015: https://www.mainstreet.com/article/from-diversity-to-dogs-the-top-10-workplace-trends-for-2015

Four Trends for the Future of Leadership Development: http://insights.ccl.org/wp-content/uploads/2015/04/futureTrends.pdf

10 Workplace Trends You'll See In 2016: http://www.forbes.com/sites/danschawbel/2015/11/01/10-workplace-trends-for-2016/

The Millennial Leadership Survey: https://workplacetrends.com/the-millennial-leadership-survey/

10 skills you need to survive the job market: http://www.theguardian.com/careers/careers-blog/ten-skills-future-work

Future Work Skills 2020: http://cdn.theatlantic.com/static/front//docs/sponsored/phoenix/future_work_skills_2020.pdf

10 C-Suite Jobs Of The Future | Fast Company | Business + Innovation: http://www.fastcompany.com/3048398/the-future-of-work/10-c-suite-jobs-of-the-future?partner=rss&utm_medium=referral&utm_source=pulsenews

American scientist Dr John Corbett uses geo-analytics and big data to help guide on-farm decision making: http://www.abc.net.au/news/2015-07-03/farming-with-big-data/6591058

BOOM: The crane industry is sending an exciting signal about the US economy: http://www.businessinsider.com/crane-industry-activity-2015-4

United Rentals : Capitalizing for the Future: A Q&A with United Rentals co-founder John Milne | 4-Traders: http://www.4-traders.com/UNITED-RENTALS-INC-14759/news/United-Rentals--Capitalizing-for-the-Future-A-QA-with-United-Rentals-co-founder-John-Milne-19948519/

Why we shouldn't get too excited about using big data for development: http://www.theguardian.com/global-development-professionals-network/2015/jun/22/why-we-shouldnt-get-too-excited-about-using-big-data-for-development

Big Data Brings Big Changes to Recruiting: http://www.wired.com/insights/2014/01/big-data-brings-big-changes-recruiting/

Big Data is Changing the Game for Human-Resource Hirers: http://mashable.com/2014/06/11/big-data-recruiting/#oCF8THwMusqw

Microsoft's Cortana Analytics looks to democratize big data: http://www.computerworld.com/article/2947612/business-intelligence/microsofts-cortana-analytics-looks-to-democratize-big-data.html

Big Data Needs Democratization, CloudSigma Says: http://www.informationweek.com/big-data/big-data-analytics/big-data-needs-democratization-cloudsigma-says/d/d-id/1297297

Big data for development: Facts and figures: http://www.scidev.net/index.cfm?originalUrl=global/data/feature/big-data-for-development-facts-and-figures.html

Big Data Could Create an Era of Big Discrimination: http://finance.yahoo.com/blogs/the-exchange/big-data-could-create-era-big-discrimination-191444085.html

How Big Data is Transforming the World of Finance: http://blog.syncsort.com/2014/09/big-data/big-data-transforming-world-finance/

Police Departments Use Big Data To Predict Where Crime Will Hit Next: http://thinkprogress.org/justice/2015/02/12/3622235/police-departments-use-big-data-predict-crime-will-hit-next/

The Big Data revolution: How data-driven design is transforming project planning: http://www.bdcnetwork.com/big-data-revolution-how-data-driven-design-transforming-project-planning

Disrupting the Construction Industry With Data Science: https://generalassemb.ly/blog/disrupting-the-construction-industry-with-data-science/

The 5 worst Big Data privacy risks (and how to guard against them): http://www.csoonline.com/article/2855641/big-data-security/the-5-worst-big-data-privacy-risks-and-how-to-guard-against-them.html

New overtime rules may have limited impact: http://www.cnbc.com/2015/06/30/new-overtime-rules-may-have-limited-impact.html

Who needs humans?: Komatsu to develop fully autonomous construction machinery: http://asia.nikkei.com/Business/Deals/Komatsu-to-develop-fully-autonomous-construction-machinery

Komatsu's Autonomous Haulage System passes 330 million tonne landmark: https://australianmining.com.au/News/Komatsu-s-Autonomous-Haulage-System-passes-330-mil

Coming to an office near you: http://www.economist.com/news/leaders/21594298-effect-todays-technology-tomorrows-jobs-will-be-immenseand-no-country-ready

Race Against The Machine: http://ebusiness.mit.edu/research/Briefs/Brynjolfsson_McAfee_Race_Against_the_Machine.pdf

As machines take on more human work, what's left for us?: http://www.pewresearch.org/fact-tank/2014/08/15/as-machines-take-on-more-human-work-whats-left-for-us/

Beyond Body Cams: Emerging Technologies That Could Help Make Police More Accountable: http://thinkprogress.org/justice/2014/12/03/3598722/technology-police-accountability/

Caterpillar, Coal Company Collaborate on Semi-Autonomous Dozers: http://www.constructionequipment.com/caterpillar-coal-company-collaborate-semi-autonomous-dozers

Cat Installs Autonomous System at Australian Mine: http://www.constructionequipment.com/cat-installs-autonomous-system-australian-mine

Are These Technologies the Future of the Construction Industry?: http://www.forconstructionpros.com/blog/12074300/are-these-technologies-the-future-of-the-construction-industry

VIDEO: Why Caterpillar's autonomous mining tech is "completely different from anything" it's ever done: http://www.equipmentworld.com/video-why-caterpillars-autonomous-mining-tech-is-completely-different-from-anything-its-ever-done/

Toward a Science of Autonomy for Physical Systems: http://cra.org/ccc/wp-content/uploads/sites/2/2015/07/Construction-v4.pdf

Toyota Will Establish New Artificial Intelligence Research and Development Company: http://www.marketwatch.com/story/toyota-will-establish-new-artificial-intelligence-research-and-development-company-2015-11-06

Google's Open Sourced TensorFlow Could Accelerate AI Research And Development: http://stgist.com/2015/11/google-open-sourced-tensorflow-could-accelerate-ai-research-and-development-5294

The Future of Employment: http://www.oxfordmartin.ox.ac.uk/downloads/academic/The_Future_of_Employment.pdf

25 Corporations Not Named Google Working On Driverless Cars: https://www.cbinsights.com/blog/autonomous-driverless-vehicles-corporations-list/

Elon Musk Donates $10M to Keep AI From Turning Evil: http://www.wired.com/2015/01/elon-musk-ai-safety/

Aerosense drones take to the skies: http://www.gizmag.com/aerosense-quadcopter-vtol-flight/39091/

Sony joins with ZMP to launch drone company: http://www.gizmag.com/sony-zmp-aerosense-autonomus-drones/38586/

How 3D Printing is Affecting the Construction Industry: http://www.forconstructionpros.com/blog/12059477/how-3d-printing-is-affecting-the-construction-industry

Big Data Europe » Empowering Communities with Data Technologies: http://www.big-data-europe.eu/

Starwood Waypoint Residential Trust to Acquire a Rival: http://www.nytimes.com/2015/09/22/business/dealbook/starwood-waypoint-residential-trust-to-acquire-a-rival.html?_r=0

Starwood Waypoint Residential Trust and Colony American Homes Announce Merger to Create Best in Class Single-Family Rental Company: http://www.businesswire.com/news/home/20150921005525/en/Starwood-Waypoint-Residential-Trust-Colony-American-Homes#.VgF9rVr3arU

Better than 50% chance recession is coming: http://www.cnbc.com/2015/09/15/better-than-50-chance-recession-is-coming-analyst.html

Tech Innovations Drive Increases in Heavy Equipment Leases and Rentals: http://www.purchasing.com/wp-content/uploads/2014/10/Tech-Innovations-Drive-Increases-in-Heavy-Equipment-Leases-and-Rentals.pdf

From Big Data to Smart Data: http://www.siemens.com/innovation/en/home/pictures-of-the-future/digitalization-and-software/from-big-data-to-smart-data-multichannel-systems.html

Amazon's data-driven approach becoming more common: http://www.newson6.com/story/29806370/amazons-data-driven-approach-becoming-more-common

Why Truck Drivers Could Be The First Casualties Of The Robot Age: http://www.fastcoexist.com/3047735/why-truck-drivers-could-be-the-first-casualties-of-the-robot-age?partner=rss&utm_medium=referral&utm_source=pulsenews

Identifying Breakthrough Life Science Research Technologies: https://www.whitehouse.gov/node/299081

NOTES

Made in the USA
Monee, IL
05 November 2026

67385917R00064